YOUR KNOWLEDGE HAS VALUE

Kissa Rajif Alunga

Methane to Liquid Fuels and Chemicals. A Catalytic Approach to Energy Sustainability and Green Chemistry

GRIN Publishing

Bibliographic information published by the German National Library:

The German National Library lists this publication in the National Bibliography;
detailed bibliographic data are available on the Internet at http://dnb.dnb.de .

Imprint:

Copyright © 2014 GRIN Verlag GmbH
Print and binding: Books on Demand GmbH, Norderstedt Germany
ISBN: 978-3-656-83026-9

This book at GRIN:

http://www.grin.com/en/e-book/282978/methane-to-liquid-fuels-and-chemicals-a-
catalytic-approach-to-energy-sustainability

GRIN - Your knowledge has value

Since its foundation in 1998, GRIN has specialized in publishing academic texts by students, college teachers and other academics as e-book and printed book. The website www.grin.com is an ideal platform for presenting term papers, final papers, scientific essays, dissertations and specialist books.

Visit us on the internet:

http://www.grin.com/

http://www.facebook.com/grincom

http://www.twitter.com/grin_com

Methane to Liquid Fuels and chemicals
A Catalytic Approach to Energy Sustainability and Green Chemistry

Kissa Rajif Alunga

College of Chemistry and Chemical engineering
Xiamen University Xiamen,
Fujian 361005, P.R China

Abstract

Methane is considered to be an alternative for crude oil in the production of petrochemicals and clean liquid fuels. It is the most abundant energy resource ever discovered in the history of petrochemical industry, often found in remote regions and serves as a feedstock for the production of chemicals and source of energy in the 21st century. Although methane is currently being used in such important applications, its potential for the production of products such as ethylene or liquid hydrocarbon fuels has not been fully realized. A number of catalytic strategies are being explored to effectively transform methane into vital products whilst considering environmental impacts, these strategies include: steam and carbon dioxide reforming, partial oxidation of methane to form CO and H_2, the direct oxidation of methane to methanol and formaldehyde, oxidative coupling of methane to ethylene and direct conversion to aromatics. Extensive utilization of methane for the production of fuels and chemicals appears to be near, but current economic uncertainties and technological associated challenges limit the amount of research activity and the implementation of emerging technologies.

Key words; Catalytic Conversion; Methane; Synthesis gas; Fuels; Petrochemicals

Table of Contents

Abstract i

1.0 INTRODUCTION 3

2.0 METHODS FOR METHANE UTILIZATION 4

    2.1.0 Indirect conversion methods 4

        2.1.1 Synthesis gas production 4

        2.1.2 Methane to liquid fuels via Fischer-Tropsch Synthesis 7

        2.1.3 Synthesis of methanol 8

    2.2.0 Direct conversion of methane into fuels and chemicals 10

        2.2.1 Conversion of greenhouse gases directly into liquid fuels 10

        2.2.2 Methane to ethylene via oxidative coupling 10

        2.2.3 Methane conversion with carbon dioxide in plasma-catalytic system 13

        2.2.4.0 Synthesis of oxygenates (methanol and formaldehyde) 14

        2.2.4.1 Methanol synthesis 14

        2.2.4.2 Formaldehyde formation 16

3.0 CHALLENGES 17

4.0 CONCLUSION 18

5.0 REFERENCES 19

1.0 INTRODUCTION

Currently catalysis is not only playing an important role in the current chemical industry for the production of key intermediates such as ketones, alcohols, epoxides among others, but is also contributing to the establishment of greener sustainable chemical processes (Zhen Guo, 2014). Concerns related to increasing prices of petroleum oil, increasingly stringent environmental legislation and dwindling crude-oil supplies mean that there is an urgent need to develop clean alternative routes to fuels and chemicals that have more readily available feedstocks.

Recent developments in natural gas production technology have led to lower prices for methane and renewed interest in converting methane to higher value products. Methane, which is the principal component of most natural gas reserves, is currently being used for home and industrial heating purposes as well as for the generation of electrical power, among others.

However, on the other hand methane is to a greater extent an underutilized resource for chemicals and liquid fuels and yet its current known reserves are enormous and more reserves continue to be discovered while crude oil deposits are dwindling.

Much of methane is produced offshore in regions far away from industrial complexes, this would require long pipeline systems to connect it to potential markets and liquefaction for shipping is expensive. Approximately 11% of this gas is re-injected, and unfortunately, another 4% is flared or vented, which is a waste of a hydrocarbon resource. Both methane itself and carbon dioxide derived from methane are greenhouse gases (Lunsford, 2000), therefore contribute to air pollution problem.

New strategies have been adopted to constructively utilize this methane through its conversion either to a transportation fuel or to a limited number of high volume chemicals such as methanol or ethylene. Here below both direct and indirect processes for converting natural gas to fuels and chemicals are discussed.

2.0 METHODS FOR METHANE UTILIZATION

It is highly desirable that methane can be utilized to produce valuable products in particular liquid fuels.

2.1.0 Indirect conversion methods

2.1.1 Synthesis gas production

Converting natural gas into a mixture of H_2 and CO, known as synthesis gas or syngas, is an important intermediate step in many existing and emerging energy conversion technologies.

Of recent, the commercialized processes that utilize methane involve an indirect process in which methane is reacted with carbondioxide (dry reforming) or with water (steam reforming) or with oxygen (partial oxidation) to form synthesis gas, which is subsequently reacted to form fuels and chemicals. The production of CO and H_2 in the appropriate ratios is achieved through the former processes.

Currently, most syngas is widely produced by steam reforming (SR) process for methanol synthesis, it is an endothermic reaction, generally represented by;

$$CH_4 + H_2O \rightarrow CO + 3H_2 \quad \Delta H = +206 \, kJ/mole$$

Methane is contacted with steam over a heated catalyst at elevated pressures and temperatures favoring syngas production at equilibrium. This process however is challenged by the demand for large initial investment for large heat exchange reactors and the problem of catalyst coking.

Catalytic partial oxidation (CPOX) of methane is a relatively inexpensive alternative method for syngas production. In the CPOX process (see fig.1), generally represented by equation below, methane is reacted with oxygen over a catalyst bed yielding syngas with an appropriate ratio$\left(\frac{H_2}{CO} \approx 2\right)$.

$$CH_4 + 1/2O_2 \rightarrow CO + 2H_2 \quad \Delta H = -38 \, kJ/mole$$

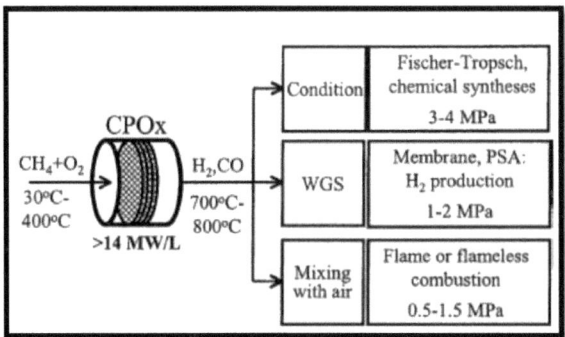

Fig.1 Routes for natural gas utilization opened or facilitated by conversion of natural gas into syngas in a high power density CPOX rector. Adapted from (M. Lyubovsky, 2005).

To avoid the energy losses associated with compression of hot, high hydrogen content, gases in the CPOX reactor should be operated at pressures in the range of 0.5 to 4 Mpa compatible with downstream processes (fig.1).

Lyubovsky et al (2005) demonstrated the operation of a methane CPOX reactor at steady state and pressures about 0.8 MPa with near equilibrium conversion of methane to above 90% and process selectivity about 90%.

Dry reforming is a catalytic process that simultaneously converts hydrocarbons to syngas using carbondioxide as an oxidizing agent, this process is highly endothermic and occurs at very high temperatures (> 700°C) to enable complete methane conversion to syngas. An added advantage of this process over steam or partial oxidation of methane is using two greenhouse gases (CO_2 and CH_4) (S.Sankaranarayanan, 2012) relieving the environment of the effects of air pollution.

$$CO_2 + CH_4 \leftrightarrow 2CO + 2H_2 \ \Delta H^{\circ}_{298k} = +247 \text{ kJ/mol.}$$

Therefore for effective conversion, selection for a suitable catalyst that can withstand high temperatures and produces high yields of syngas is significant for this reaction. Ni has been most effectively used due to its lower costs and easier availability than noble gases such as Rh, Rn among others. However it is affected by the catalyst coking that renders it inactive.

Catalytic partial oxidation, with the use of nearly pure O_2, is said to circumvent the coking problem (Lunsford, 2000).

The support catalysts and promoters have been reported to play a crucial role in the dry reforming reaction. Of recent Labrecque et al (2011), investigated the activity of Fe catalyst in the presence of electrical current and found that electron flow and addition of water vapour play a crucial role in methane conversion. They showed that hot water saturation of the CO_2 and CH_4 mixture permitted excellent conversion of methane to syngas with a composition very close to the expected equilibrium calculations.

Carbon Sciences Inc., a company developing a technology to transform CO_2 and CH_4 into gasoline and other portable fuels has been developing its own catalyst for the efficient transformation of CO_2 and CH_4 gas into synthesis gas, which can then be further catalytically processed into gasoline and other fuels (see fig.2).

In December 28th 2010, University of Southampton (UoS) and Carbon Sciences Inc. announced signing a worldwide exclusive license agreement for the patented technology for the dry reforming of methane with CO_2.

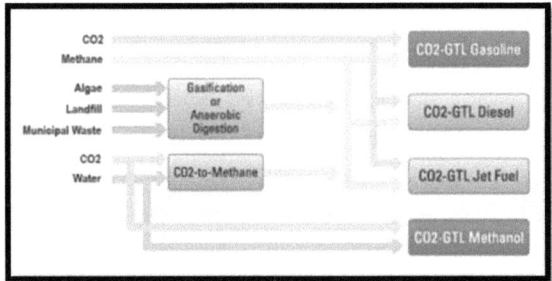

Fig. 2, Carbon Sciences Process Block Diagram. Adapted from (new energy and fuel, 2011)

2.1.2 Methane to liquid fuels via Fischer-Tropsch Synthesis

Fischer-Tropsch synthesis (FTS) demonstrates a great capability for production of clean transportation fuels, chemicals, among other products via transformation of syngas. FTS involves the hydrogenation of carbon monoxide into higher hydrocarbons especially higher paraffins.

$$nCO + 2nH_2 \rightarrow -[CH_2]_n - +nH_2O \; \Delta H^0_{298} = -204kj/molCO \ldots\ldots.1$$

$$nCO + 2nH_2 \rightarrow -[CH_2]_n - +nCO_2 \; \Delta H^0_{298} = -166kj/molCO \ldots\ldots 2$$

The stoichiometry 2 is more suitable for a particular catalysts with high water gas shift activities like iron catalyst.

All the group VIII metals exhibit high activity for carbon monoxide hydrogenation with different selectivities. Among them Co, Fe and Ru have been shown to display high activity to formation of higher hydrocarbons. Due to their high activity for FTS, high selectivity to linear products, more stability toward deactivation, low activity toward water-gas shift reaction (WGSR) and low cost compared to Ru-based catalysts, cobalt-based catalysts are the preferred choice for FTS.

Tsubaki et al (2014) studied FTS performance of a highly dispersed Co/SiO_2 catalyst prepared by citric acid synthesis in argon atmosphere. In their report, it was shown that the catalyst activity was about four-times higher than that of the catalyst prepared by conventional incipient-wetness impregnation method.

Supercritical solvents play a critical role in chemical reactions and FTS because of their single-phase operation with densities sufficient enough to enable substantial dissolution power, whilst enhancing diffusivities that are higher than those of normal liquids and viscosities lower than their liquid counterparts. Ma et al. (2014) demonstrated the performance of catalyst composed of Cu–ZnO and Pd/ZSM 5 in a near-critical n-hexane solvent for syngas conversion to gasoline. Their results showed a decrease in the particle size of ZSM 5 and an increase in the Pd loading leading to selective production of hydrocarbons in the gasoline fraction.

2.1.3 Synthesis of methanol

It is anticipated that the requirement for methanol in an expanding world market will be produced primarily from natural gas in the near future. Currently methanol is produced via multi-step process i.e. methane-syngas-methanol, which process involves high capital costs (N. V. Beznis, 2010). However, a non-energy-intensive, one step process for selective oxidation of methane to methanol would be preferred on industrial scale as we shall see under direct methods.

N. V. Beznis et al (2010), investigated a series of Cu-ZSM-5 zeolites (prepared by varying nature of the charge compensating cation, copper precursor, copper loading and pH) while varying different copper species on methane oxidation to methanol. In this study it was demonstrated that Cu-ZSM-5 zeolites was capable of converting methane directly to methanol using oxygen as the oxidant. Beznis's team established two kinds of copper Cu-O, one on the outer surface of the catalyst which was inactive for methanol production and the one inside the channels of Cu-ZSM-5 zeolite responsible for the selective conversion of methane to methanol.

Hammond et al (2012) demonstrated a new low-temperature route that produces methanol at high catalytic rates (TOF[1] >14 000 h^{-1}) and with good selectivity ($\leq$95%) under mild and environmentally benign reaction conditions ($\leq$50°C, aqueous medium). In their experiment, they showed that Cu-Fe/ZSM-5(30)/ H_2O_2 can act as an efficient heterogeneous catalyst for the selective oxidation of methane under mild reaction conditions. The catalyst is responsible for the activation of both the substrate and the oxidant, and plays an intimate role in all aspects of the catalytic cycle.

Periana et al (1998) developed various oxidation catalysts based on Pt(II), Pd(II), and Hg(II) salts that have been proven to functionalize C–H bonds, consequently resulting into better yields of partially oxidized products (Eq.3). For instance, Pariana and team demonstrated selective oxidation of methane in concentrated sulphuric acid to give methyl bisulphate at temperatures around 473K over [(2,2_-bipyrimidine)PtCl$_2$] catalyst (see Fig.3), they were able to achieve 81% selectivity to methyl bisulfate, a methanol derivative, which was then hydrolyzed to methanol (eq.4)

---

[1] TOF stands for turn over frequency which describes catalyst activity.

$$CH_2 + 2H_2SO_2 \rightarrow CH_3OSO_3 + SO_2 + 2H_2O........3$$

$$CH_3OSO_3 + H_2O \rightarrow CH_3OH + H_2SO_4...............4$$

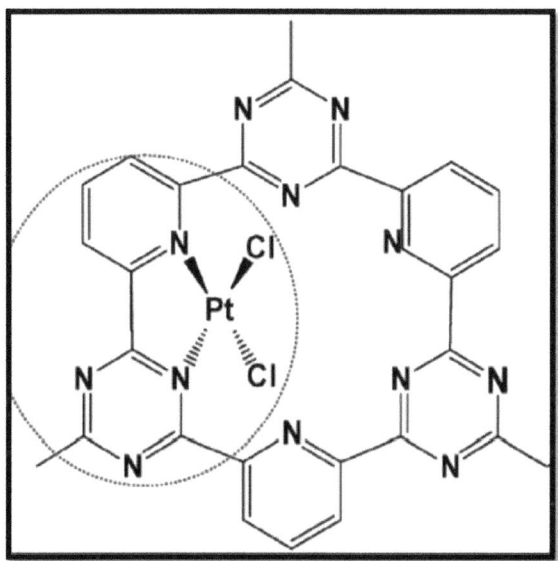

Fig. 3.Bipyrimidyl Pt(II) complex used in the oxidation of methane to methylbisulfate in concentrated sulphuric acid. Adapted from (M.C. Alvarez-Galvan, 2011)

2.2.0 Direct conversion of methane into fuels and chemicals

The direct methods for methane conversion into high value products have an economic advantage over indirect methods, however low yields associated with direct methods have limited progress into commercialization stages.

2.2.1 Conversion of greenhouse gases directly into liquid fuels

Methane and carbondioxide are greenhouse gases whose increase in emissions have made significant climate changes such as increase in global temperatures and changes in wind patterns. Of recent, a lot of research is being undertaken to convert methane and carbondioxide directly into liquid fuels as a strategy to minimize carbon emissions, there are claims that such technology has been discovered but so far little attempts have succeeded in this area.

Liviu M. Mirica, PhD, assistant professor of chemistry at Washington University in St. Louis may have found and is developing a novel metal catalyst that would be able to turn greenhouse gases like methane and carbon dioxide into liquid fuels without producing more carbon waste in the process. Mirica describes a new metal complex that can combine methyl groups (CH_3) in the presence of oxygen to produce ethane (CH_3-CH_3), the new catalyst combines methyl groups (CH_3) molecules in the presence of oxygen to produce ethane, this is the second step in the conversion of methane (CH_4) into a longer-chain hydrocarbon, or liquid fuel.

2.2.2 Methane to ethylene via oxidative coupling

Ethylene is the most widely used chemical in the world as well as an important natural plant hormone, used in agriculture to enhance fruit ripening. With the declining crude oil reserves, definitely ethylene production will follow suit, this has prompted the industrial sector to seek for alternative sources and production methods for ethylene such as direct oxidative coupling of methane (OCM), and it involves reaction of CH_4 and O_2 over a catalyst at elevated temperatures (600-800°C) to form C_2H_6 as a primary product and C_2H_4 as a secondary product.

Despite its profitability potential as a new production technique, it is challenged by low C_2 yields and separation requirements. Secondly, the unselective gas-phase radical reactions have been reported to be of detrimental influence to OCM because of deep oxidation of C_2 products to COx

(C0 and CO_2), furthermore the stability and deactivation of the catalysts are still a major problem such as nickel based catalysts (J. B. Branco, 2014).

However, J. B Branco et al (2014) have shown that strong acidic inorganic compounds are among the most active, selective and stable catalysts for the activation of CH_4 using N_2O as oxidant, use of N_2O rather than O_2 as an oxidant agent for the activation of methane justifies the development of an environmentally benign solution since N_2O is recognized as a greenhouse gas.

On the other hand, applications of molten salts such as acidic alkali molten metal chlorides, are well recognized catalysts, they can also act as active media that disperse other catalytic active complexes. Molten salt catalytic properties are mainly linked to the high mobility of cations and anions within the melt. Therefore, molten salts can be used as an attractive and viable alternative to the classical catalysts used for the OCM process.

Furthermore, J. B Branco et al studied the catalytic behavior of bimetallic oxides containing f block elements using binary intermetallic compounds $LnCu_2$ (Ln = La, Ce, Pr, Eu, Gd, Dy, Tm) LnNi (Ln=Pr, Gd, Lu), $ThCu_2$ and $AnNi_2$ (An=Th, U) as catalysts precursors. These compounds exhibited activity and selectivity for the partial oxidation of methane for the production of light hydrocarbons.

In their study of the catalytic oxidation of methane over acidic $KCl-LnCl_3$ (Ln=La, Ce, Sm, Dy and Yb) molten salts using molecular oxygen as oxidant, confirmed that the potassium-lanthanide chloride molten salts are active and remarkably selective for the production of C_2 hydrocarbons, $KCl-LnCl_3$ (Ln = La, Ce, Sm, Dy and Yb) molten salts were active and selective for the activation of methane using nitrous oxide as oxidant. They observed reaction activity increase with the temperature and the reaction main products were C_2 hydrocarbons (ethane and ethylene) and CO, CO_2. At 750 °C the desired high conversion of methane and higher production of C_2 hydrocarbons was realized. At such temperature, all catalysts are highly selective to C_2 hydrocarbons (conversion 10–20%, selectivity 70–80%).

In their conclusion, Potassium-lanthanide chloride molten salts ($KCl- LnCl_3$, Ln = La, Ce, Sm, Dy and Yb) were used as catalysts for the oxidation of methane with N_2O. They proved to be active and selective for the production of C_2 hydrocarbons for example CH_4 conversion $\approx 15\%$

and C_2 selectivity of 75% (see fig.4) over the $KCl–CeCl_3$ catalyst at 750 °C and remarkably stable for long periods of time on the gaseous stream.

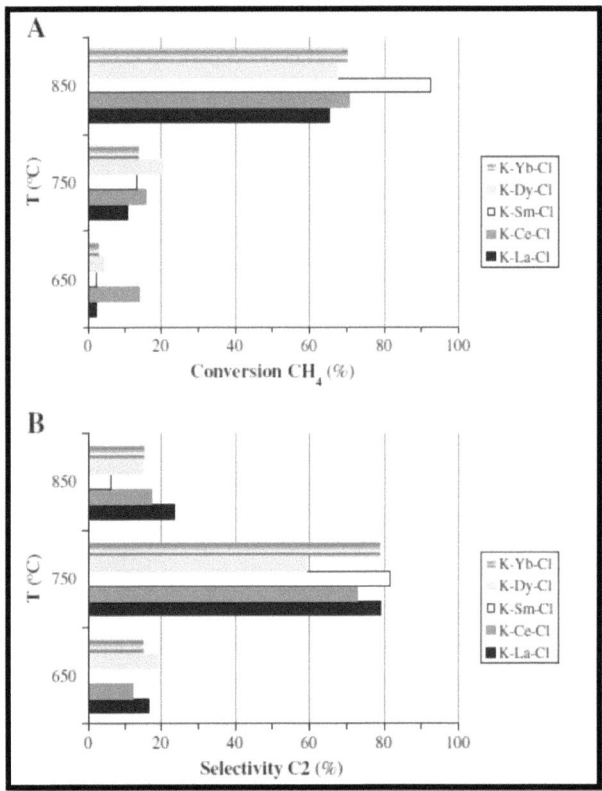

Fig.4. Effect of the temperature on the conversion of methane (A) and selectivity to C_2 hydrocarbons (B) over $KCl–LnCl_3$ (Ln = La, Ce, Sm, Dy and Yb) molten salts. Adapted from (J. B. Branco, 2014).

In the area of oxidative dehydrogenation of ethane to ethylene (ODHE), Zhu et al. (2014) applied the sol–gel method for the synthesis of Zr, Ti, Mo, W, and V modified NiO based catalysts. They found that the introduction of group IV, V and VI transition metals into NiO reduces the catalytic activity in ODHE. In turn, the best effects on ethylene yield were obtained with W–Ni–O and Ti–Ni–O catalysts (H. Zhu, 2014).

Another study of catalytic partial oxidation of ethane by Cimino and co-workers focused on Rh and Pt based honeycomb reactors in the presence of sulfur (as SO_2 up to 51 ppmv). The findings showed that overall catalytic performance was enhanced by the selective S-poisoning of catalytic steam reforming of C_2H_6 and C_2H_4 over Rh, as well as by the inhibition of C_2H_6 catalytic oxidation when H_2 was co-fed as sacrificial fuel over both Rh and Pt catalysts (S. Cimino, 2014).

2.2.3 Methane conversion with carbon dioxide in plasma-catalytic system

Recently K. Krawczyk et al (2014) investigated methane conversion to higher hydrocarbons and alcohols under oxidative conditions (with CO_2) in plasma-catalytic system (HPCS) with dielectric barrier discharge (DBD) (see fig.5), "the hybrid plasma-catalytic system (HPCS) was used for the oxidative coupling of methane (OCM) with CO_2, a dielectric barrier discharge (DBD) reactor was powered at the frequency of about 6 kHz. Molar ratio $[CO_2]$: $[CH_4]$ in the inlet gas mixtures ($CH_4 + CO_2 + Ar$) was 1" (K. Krawczyk, 2014).The solid catalysts or packings studied were iron on alumina ceramics and zeolites (NaY and Na-ZSM-5).

In their findings, the products were Hydrogen, CO, hydrocarbons ($C_2 - C_4$), and alcohols found in the outlet gas and the conversion of methane to these compounds was the highest with Na-ZSM-5 zeolites, the dominant organic products included ethane and propane.

In addition, alcohol compounds (ethanol and methanol) were obtained in a homogeneous system and with a carrier (Al_2O_3), among the two catalysts system, formation of alcohols was higher with Fe/Al_2O_3 catalyst than that with Na-ZSM-5 and NaY.

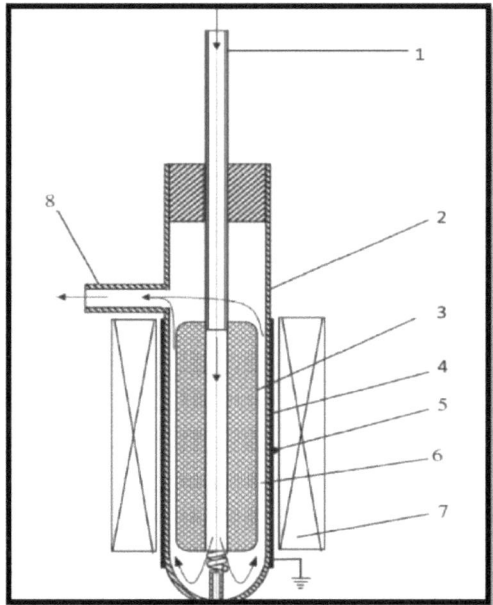

Fig.5. Reactor for methane conversion in plasma-catalytic system. 1-gas inlet, 2-quartz tube, 3-high voltage electrode, 4-grounded electrode, 5-thermocouple,6-discharge gap (3 mm), 7-heater, 8-gas outlet, → −direction of gas flow through the reactor. Adapted from (K. Krawczyk, 2014)

Conclusively, plasma-catalytic system (HPCS) showed an overall improvement in the conversion of methane and carbon dioxide to higher hydrocarbons and alcohols.

2.2.4.0 Synthesis of oxygenates (methanol and formaldehyde)

A direct route for the synthesis of methanol and formaldehyde has proved to be one of such greatest achievements among methane conversion technologies.

2.2.4.1 Methanol synthesis

Developing a direct route for conversion of methane to methanol can provide the foundation for efficient utilization of natural gas. Methanol (CH_3OH) is an energy-dense liquid that can be transported easily with existing infrastructure.

In addition, it is a versatile molecule as it can be used for fuel cells, blended with gasoline, converted to gasoline or dimethyl ether which is a component of diesel fuel and converted to ethylene and propylene, which are precursors to a wide range of chemicals. Given the chemical properties and versatility of methanol as well as the abundance of natural gas, the direct conversion of methane to methanol is indeed a dream process that we need to achieve.

The direct oxidation of methane to methanol is an exothermic process as indicated by the subsequent reaction equation.

$$CO_4 + 0.5O_2 \rightarrow CH_3OH \quad \Delta H_{298}^0 = -126kj/mol$$

An in situ direct conversion of methane to methanol has been proposed using the ZSM-5 supported copper oxide nanocatalyst and a high selectivity and high yield have been reported as well i.e. 100% and 31.5% respectively (M. Gharibi, 2012).

Another new class of solid catalyst has of recent been published for the direct low-temperature oxidation of methane to methanol. This catalyst is a covalent triazine-based framework (CTF) with various accessible bipyridyl structure units, which should allow the coordination of platinum, and resemble the sites for platinum coordination in the molecular Periana catalyst (See fig.6).

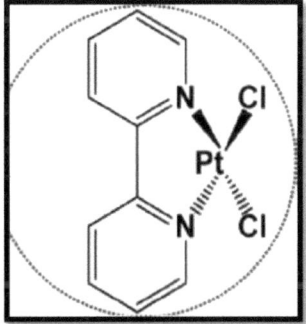

Fig.6. Covalent triazine-based framework (CTF) with numerous accessible bipyridyl structure units which are suitable to coordinate Pt(II) complex. Adapted from (M.C. Alvarez-Galvan, 2011)

There has been a promising progress in biocatalysis where enzyme monooxygenase converts methane to methanol at ambient conditions (temperature and pressure) using iron centers with the aid of NADH as a reductant. Following the success of enzyme reaction, $FePO_4$ has been shown as the principle model catalyst for production of methanol from methane and O_2 reagents when H_2 is added (Lunsford, 2000). In the account of the reaction, H_2 reacts with O_2 forming surfaces responsible for methane activation.

Despite all the progress made in the direct conversion of methane to methanol, full scale commercialized stage has not yet been achieved.

2.2.4.2 Formaldehyde formation

Formaldehyde may be formed at high temperature oxidation of methane over pure SiO_2 and V_2O_5/SiO_5, however this reaction route has not significantly broken through in the single-pass yield i.e. 3-4% yield of HCHO is realized.

Other studies have reported successful preparation of highly selective catalyst for HCHO production. It has been demonstrated that active phases such as MoOx, VOx and $FePO_4$ greatly enhanced catalytic performance of formaldehyde formation through heterogeneous direct oxidation of methane. In addition, Wang et al, confirmed that $FePO_4$ nanoclusters confined in the mesoporous channels of MCM-41 or SBA-15 exhibited higher activity and selectivity of HCHO formation than the crystalline $FePO_4$ (Y. Wang, 2004).

3.0 CHALLENGES

However, although there has been significant progress in the catalytic conversion of methane to high value products, there are still numerous setbacks curtailing natural gas conversion technologies, here a few of them are discussed.

Catalysts involved in direct methane conversion methods produce only a few desired products such as CO/H_2, acetylene, ethylene, aromatics, methanol, HCHO. Secondly, methane reaction products are much more reactive than methane itself, their reaction lead to thermodynamically more stable carbon or CO_2 products.

In some reactions, product yields are quite low, realization of high yields requires continuous removal of desired products from the reaction zone and kinetic guarding against secondary reactions.

The rapid subsequent combustion reactions of methanol to form CO_2 limit conversions in practice whereas the explosive nature of necessary reacting mixtures creates challenges in the design of mixing schemes and pressure vessels.

Biocatalysts' selective conversion of methane to methanol has been purported as an ultimate replacement for other chemical routes, however this will need enzymes that perform selectively and efficiently in concentrated aqueous methanol solutions.

Methanol formation via indirect routes that involve Hg complexes and sulphuric acid are very difficult to handle due toxicity associated with them.

4.0 CONCLUSION

The development of conversion technologies of methane into useful chemicals and fuels are currently one of the hot topics in the field of research and development given the fact that crude oil prices are soaring as well as declining oil reserves.

With natural gas currently priced in the U.S. at less than 25% the energy value of oil, wouldn't there be a huge opportunity to reform CO_2 and methane into more useful chemicals and fuel? Therefore it would be a general economy growth stimulus if this energy resource is farther developed, thus providing cheaper fuel supply.

There are a number of processes that have been developed to enhance methane utilization, these are broadly divided into two groups; indirect and direct processes as have been discussed. The indirect processes rely upon the formation of synthesis gas (CO and H_2) either by reforming reactions or by partial oxidation. In the direct processes, methane may be converted to methanol, formaldehyde, ethylene or aromatics.

Catalyst developments have always played a crucial role in the industrial processing plants and they have been at the forefront in the advanced energy processing technologies, nonetheless, there remains challenges as discussed earlier especially achieving higher selectivity by a catalyst to the desired product and developing a catalyst tolerant to adverse conditions as well as having longer life. Therefore, successful methane conversion will require catalyst innovations and novel engineering developments.

5.0 REFERENCES

C. Hammond, R. L. (2012). Catalytic and Mechanistic Insights of the Low-Temperature Selective Oxidation of Methane over Cu-Promoted Fe-ZSM-5. *Chem. Eur. J.*, *18*, 15735 – 15745.

H. Zhu, H. D. (2014). Metal oxides modified NiO catalysts for oxidative dehydrogenation of ethane to ethylene ethane to ethylene. *Catalysis Today*, *228*, 58–64.

J. B. Branco, A. C. (2014). Oxidative coupling of methane over KCl–LnCl$_3$ eutectic molten salt catalysts. *Journal of Molecular Liquids*, *191*, 100–106.

J.J. Spivey, G. H. (2014). Catalytic aromatization of methane. *Chem. Soc. Rev.*, *43*, 792-803.

K. Krawczyk, M. M.-S. (2014). Methane conversion with carbon dioxide in plasma-catalytic system. *Fuel*, *117*, 608–617.

L. Shia, C. Z. (2014). Citric acid assisted one-step synthesis of highly dispersed metallic CO/SiO$_2$ without further reduction: As-prepared CO/SiO$_2$ catalysts for Fischer–Tropsch synthesis. *Cat. Today*, *228*, 167–174.

Labrecque R, L. J. (2011). Dry reforming of methane with carbondioxide on an electron-activated iron catalytic bed. *Bioresource technology*, *102*, 11244-11248.

Lunsford, J. H. (2000). Catalytic conversion of methane to more useful chemicals and fuels: a challenge for the 21st century. *Catalysis Today*, *63*, 165–174.

M. Gharibi, F. Z. (2012). Nanocatalysts for conversion of natural gas to liquid fuels and petrochemical feedstocks. *Applied Catalysis A: General*, *443– 444*, 8– 26.

M. Lyubovsky, S. R. (2005). Catalytic partial "oxidation of methane to syngas" at elevated pressures. *Catalysis Letters*, *99*.

M. Zabeti, W. M. (2009). Activity of solid catalysts for biodiesel production: A review. *Fuel Processing Technology*, *90*, 770-777.

M.C. Alvarez-Galvan, N. M. (2011). Direct methane conversion routes to chemicals and fuels. *Catalysis Today*, *171*, 15-23.

N. V. Beznis, B. M. (2010). Cu-ZSM-5 Zeolites for the Formation of Methanol from Methane and Oxygen: Probing the Active Sites and Spectator Species. *Catal Lett*, *138*, 14–22.

new energy and fuel. (n.d.). Retrieved June 16, 2014, from http://newenergyandfuel.com/http:/newenergyandfuel/com/2011/01/03/make-liquid-fuels-from-methane-and-CO$_2$/

R. A. Periana, D. J. (1998). Platinum Catalysts for the High-Yield Oxidation of Methane to a Methanol Derivative. *Science , 280*, 560-564 .

S. Cimino, G. M. (2014). Ethane catalytic partial oxidation to ethylene with sulphur and hydrogen addition over Rh and Pt honeycombs. *Catal. Today , 228*, 131–137.

S.Sankaranarayanan, K. (2012). Carbondioxide-A potential raw material for the production of fuel, fuel additives and bio-derived chemicals. *Indian Journal of Chemistry , 51A*, 1252-1262.

T. Ma, H. I. (2014). Selective synthesis of gasoline from syngas in near-critical phase. *Catal. Today , 228*, 167–174.

Y. Wang, X. W. (2004). SBA-15-supported iron phosphate catalyst for partial oxidation of methane to formaldehyde. *Catal Today* , 155–161.

Zhen Guo, B. L. (2014). Recent advances in heterogeneous selective oxidation catalysis for sustainable chemistry. *Chem. Soc. Rev. , 43*, 3480-3524.